EYEWITNESS
EXPLORERS

Rocks and Minerals

Written by
STEVE PARKER

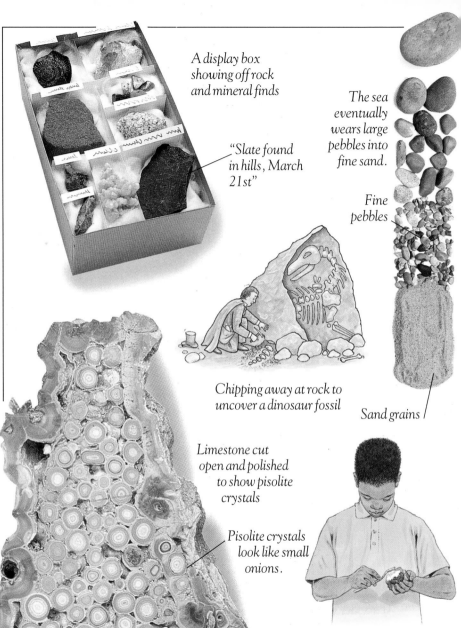

A display box showing off rock and mineral finds

"Slate found in hills, March 21st"

The sea eventually wears large pebbles into fine sand.

Fine pebbles

Chipping away at rock to uncover a dinosaur fossil

Sand grains

Limestone cut open and polished to show pisolite crystals

Pisolite crystals look like small onions.

Use an old toothbrush to clean your rocks.

EYEWITNESS
EXPLORERS

Rocks and
Minerals

Written by
STEVE PARKER

A DK PUBLISHING BOOK

Editor Djinn von Noordon
Designer Sharon Grant
Senior editor Susan McKeever
Art editor Vicky Wharton
U.S Editor Charles A. Wills
Production Catherine Semark

First Paperback Edition, 1997
2 4 6 8 10 9 7 5 3 1

Published in the United States by
DK Publishing, Inc., 95 Madison Avenue
New York, New York 10016
Visit us on the World Wide Web at
http://www.dk.com

ISBN 0-7894-1682-4

A CIP catalog record is available from the Library of Congress.

Color reproduction by Colourscan, Singapore
Printed in Spain by Artes Gráficas Toledo, S.A.
D.L. TO:8-1997

Contents

Rocks and minerals

Our world is built of rocks and minerals. They are fascinating, often beautiful, and fun to study. Rocks and minerals don't run away like animals, or die in the winter like some plants. With a little time and some basic equipment, you will soon become an expert explorer!

Out and about
You can go rock and mineral exploring almost anywhere. Try and find a new rock in every new place.

Tourmaline crystal

Minerals everywhere
These mineral crystals formed from hot watery solutions inside the Earth. Can you see how each kind of crystal grew into a different shape?

Crystals formed very slowly and gradually grew together.

Chipped edge is very sharp

Quartz crystal

Albite crystals

Useful rocks
Thousands of years ago, Stone Age people did not have metals or plastics. They made their tools and weapons out of stone. This dagger blade was carefully chipped from a lump of flint rock.

Strike it rich!

People who study rocks and minerals are called geologists. There are many different kinds of geologists. Some might find the best places to build bridges and roads; others can help discover new stores of oil, gems, and precious metals in the Earth.

Equipped to explore

You can explore the world of rocks and minerals with just your eyes, but a little equipment helps a lot. Draw pictures of the places you visit and the finds you make in a notebook.

Goggles protect your eyes when chipping away at rocks with a hammer.

You can use old newspapers to wrap up specimens you've found.

A magnifying glass will help you see rocks in detail.

Use a small notebook while out exploring. You can transfer the pages into a scrapbook when you get home.

Always wear goggles when chipping at rocks.

Careful collecting

Check with an adult before you collect rocks and minerals. A backpack is a good carrier for your best finds. Wrap them in newspaper to protect them, and to make sure they do not stick into your back.

Rock or mineral?

Look out across the land. Now imagine the layer of plants, fields, and houses disappearing. What's left underneath? Rocks! Rocks are made from materials called minerals. Hundreds of minerals, in different amounts, make up hundreds of different kinds of rocks.

Mineral mosaic

Granite is a hard, tough rock. Like all rocks, it is made of minerals. You can see gray areas of quartz, black bits of mica, and pink and white feldspars. Examples of these minerals are shown in close-up below.

Black mica

Mineral grains are big enough to see clearly

Quartz is light-colored

Gray quartz

Pink feldspar

Look for small, glittery crystals.

Milky-white or pale-pink crystals

Quartz

The sand grains you find on a beach are made from quartz. Quartz contains silicon and oxygen and is very hard.

Mica

All minerals are made from chemicals. Mica contains the chemicals aluminum, potassium, and silicon.

Feldspars

Feldspars are very common minerals. They also contain aluminum and silicon.

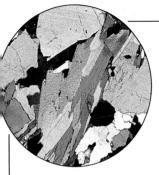

A closer look

In some rocks, the mineral grains are very small. Geologists need magnifying glasses and microscopes to see their shapes and colors. Here, a microscope shows you the minerals in a slice of quartz.

Organic minerals

Minerals aren't just found in rocks. Some minerals, called organic minerals, come from plants and animals. They include jet, coral, amber, and pearls. Jet (right) is formed over millions of years from trees that have rotted and squashed together. When carved and polished, jet looks like black glass.

Mini rock-makers

Corals are sea animals, a bit like tiny jellyfish. Each one takes minerals from the water and makes a stony cup around itself. Millions of cup "skeletons" pile up into a rocky coral reef.

Life's work

If a grain of sand gets stuck inside an oyster's shell, the oyster makes a pearl around the grain. It takes three years or more for a large pearl to form.

Trapped in time

In prehistoric times, resin (sap) from trees hardened into amber. Plants and small creatures became trapped inside – for millions of years!

Types of rocks

The Earth was formed nearly 4,600 million years ago – but rocks are not that old. Through the ages, old rocks have melted and cooled again, some have been changed by pressure and heat, and some rocks have worn away into tiny pieces. These endless changes make new rocks from old ones.

When rocks melt

Sometimes rocks and minerals heat up so much that they melt and turn into liquid rocks. When they cool down again, they harden into igneous rocks. *Igneous* means "made by fire."

Gabbro is an igneous rock with large mineral crystals.

Changed rocks

When they are squeezed hard or heated, some rocks change their form and color. Rocks made like this are called metamorphic rocks.

Liquid rock pours from volcanoes and cools into igneous rock.

This metamorphic rock is called gneiss. Can you see the bands of dark minerals?

Rocks under the mountains are squashed to make metamorphic rocks.

Strong or weak?
Rocks, minerals, and crystals are made from millions of tiny parts called molecules. If the molecules are in a neat and regular pattern, the rock is hard and firm. If they are scattered around, the rock is weak and crumbly.

Cemented rocks
Some rocks wear down into small pieces that settle in deep layers. Gradually, more layers pile on top. The layers squash and cement together to make a new kind of rock, called sedimentary rock. *Sedimentary* means "settled down."

This sedimentary rock is called quartz conglomerate.

You'll often find quartz conglomerate near rivers and beaches.

Seashore pebbles are stuck in the rock.

Ice, rain, rivers, and wind wear rocks into smaller pieces.

Rivers carry bits of rock to the sea.

Seaside mix
Tiny pieces of mud and sand sink to the bottom of the sea, along with bits of shell and coral. They form layers of sludge, mud, and ooze. Gradually, these harden into sedimentary rocks.

13

Making a collection

It's easy to start your own rock and mineral collection. Keep an eye out for unusual specimens near rocky hillsides, cliffs, and road cuttings, and explore riverbanks, lakesides, and the beach. Try looking for rocks and minerals in different colors – you'll find black, brown, yellow, white, and even pink ones.

Lots of rocks

Don't bring home every rock you see – you won't have enough room to keep them all! Take only small samples, and leave some for other people to admire.

Select from nature

A whole beach or hillside covered with rocks looks like a jumble. But if you gather a few of the best ones and clean them, you will see their interesting colors and shapes more clearly.

Choose rocks that are not cracked or broken.

Brown iron coating on pyrite nodule

Small rocks are easier to carry home.

Green glass, rounded and smoothed by the sea

Cleaning your rocks

Soak any sturdy rocks in warm soapy water. Very muddy or dirty finds will need a careful scrub with an old, wet toothbrush. Rinse them under the faucet, and leave the clean rocks to dry on a newspaper.

Make a display box

Keep your best finds on show by making a display box. You will need a shoebox, stiff cardboard, cotton batting, glue, and scissors.

Place rocks with their best side upward.

1 Cut strips of cardboard that fit lengthways and crossways into the box. Make slits in them so they slot together.

2 Glue the strips to make box-shaped sections. Divide a few sections with smaller cardboard strips. Line the sections with cotton batting.

Group rocks by their colors or by the places where you found them.

"Slate found in hills, March 21"

"Sandstone found in valley, June 3"

Where and when?
Label each find with its name, and where and when you found it.

Cotton batting protects the rocks and minerals on display.

The rocky Earth

To people on the surface, the world looks flat. But if you sliced the Earth open like an apple, you would find many kinds of rocks, all packed together in different layers like an enormous onion.

Deepest dig
Even if you dug the deepest hole you could, it would be just a tiny pinprick on the Earth's surface.

Outer core has very hot liquid rocks

Inner core is made up of solid rocks

Bump on the Earth
The tallest mountain, Mount Everest, is only a tiny bump on the thin surface of the Earth. People used to think the Earth was a solid ball – but now we know there are more layers under the crust, and not all are solid.

Thick mantle has rocks like stiff, red-hot jelly

Crust is a layer of solid rocks

In the Earth's core

A meteorite is a rock that comes from outer space. But it is made of the same iron-rich rock as the Earth's core. You will never see or touch the center of the Earth – the temperature there is hot enough to boil iron!

This meteorite has been cut, etched, and polished.

Can you see the crystal patterns?

Basalt has many tiny grains of different minerals.

Basalt crust

This rock is called basalt. Basalt is one of the most common rocks in the Earth's crust. A hard and heavy igneous rock, basalt forms huge sheets over the land and the ocean floor.

This olivine basalt comes from a volcano crater in Hawaii.

Small garnet crystals reflect the light.

Mantle movement

This eclogite rock comes from deep inside the Earth's upper mantle. It was brought to the surface during violent Earth movements. Eclogite is a metamorphic rock. It is speckled with garnet crystals and green pyroxene minerals.

Red garnet crystals

Green pyroxene minerals

Thin skin

Slice of Earth

Why is the Earth like an apple? Because they both have cores! And they both have very thin skins (or crusts). Carefully cut an apple in half and look at the skin. Compared with its size, the crust of the Earth is even thinner.

17

Moving rocks

You may feel like you're standing on solid ground, but the rocks under your feet are always on the move. The Earth's crust is made of gigantic blocks of rock called plates. Over millions of years, the plates collide and crumple to make mountains, valleys, and canyons.

200 million years ago

135 million years ago

Today

Splitting rocks
When the plates pull apart or slide past each other, an earthquake may happen.

Peaks are sharp and jagged.

Floating around
The Earth's continents lie on the plates. Like huge jigsaw pieces, the plates float on the semi-liquid rock underneath; 200 million years ago there was only one continent.

Bump and buckle
As the Earth's plates move around, they bump into each other. Their edges slowly buckle. Fold mountains such as the Himalayas are made like this.

Layers of lava

When liquid rock – called lava – erupts from a volcano, it flows away, cools, and hardens. After many eruptions, layers of lava build up into volcanic mountains.

Long, sloping sides are made from many layers of cooled lava flows.

Block mountain

Slip away

When huge rock blocks tilt or drift apart, the land between them slips and falls into the rift (gap) between the rocks. The highlands become block mountains, and the lowland becomes a rift valley.

Split between the rocks makes the new rift valley

Move a mountain

You can pretend to crush a massive block of rocks into a range of mountains – and all you need is modeling clay! Choose two or more colors of clay for the different layers of rock.

1 Squash the layers of clay on top of each other. Hold the block at each end – now you have the flatlands.

2 Squeeze your hands together. The layers bow and buckle into humps or mountains.

Volcanic rocks

Deep in the Earth's crust, rocks get so hot that they melt. Molten rock that reaches the surface is called lava. Molten rock underground is known as magma. And as lava and magma cool and harden, they form volcanic rocks of fantastic shapes and sizes.

Stretchy skin

When the lava surface hardens into a crust, the molten lava underneath keeps on flowing. It twists the crust into lumpy, rope like shapes – that's why this lava is called ropy lava.

You can walk on the crust if it is thick enough, with the red-hot lava flowing underneath!

Black glass

This rock is called obsidian. It was made by lava that cooled too fast for crystals to grow. It is smooth, glossy, and sharp, like black glass.

Rivers of rock

Molten lava is like a red-hot river of rock, setting fire to everything it touches. The crust on this lava has already started cooling. Can you see the glowing liquid rock underneath?

This hole is called the vent.

Lava layers from old eruptions.

Inside a volcano

Most volcanoes are cone-shaped mountains made of many layers of cooled hardened lava. Each time this kind of volcano erupts, runny lava spurts out of the hole, flows down the sides, and then becomes solid.

Lava oozes from side vents

The main magma chamber is deep underground.

Make a volcano

Take the cap off a bottle of fizzy water, and make your own erupting volcano! You'll need some red food coloring and a plastic bottle of fizzy water.

Old Devil

Devil's Tower, in Wyoming, is made from lava that hardened inside a volcano. The softer rock of the volcano itself has worn away.

Plastic bottle of fizzy water

1 Add red food coloring to make the water look like red-hot lava.

2 Tighten the cap. Gently shake the bottle. Hold it away from your face.

3 Unscrew the cap to release the pressure. The water erupts like a volcano!

Gas bubbles get bigger and make the water frothy.

21

Igneous rocks

Do you know how hot boiling water is? A steaming cup of tea or coffee can burn you badly if you spill it on your skin. Some rocks deep in the Earth get ten times as hot as boiling water and melt. When these melted rocks cool, they form igneous rocks. If they cool quickly, they make igneous rocks with small crystals. If they cool slowly, over thousands of years, they make igneous rocks with larger crystals.

Floating on water
Would you believe that a rock can float? Pumice is hardened lava froth. Just like the froth on top of soda, pumice is packed full of tiny air bubbles. The air trapped inside keeps the stone floating on water.

Pele's hair is made of basalt. It also contains tiny crystals of a mineral called olivine.

Fine strands of glasslike rock look just like hair.

Rocky hair
When an igneous rock hardens in air, on the Earth's surface, it is called an extrusive igneous rock. One example is Pele's hair. It forms when an erupting volcano sprays fine streams of lava into the air. The lava cools and hardens almost at once to form an amazing "hairy" rock.

Buried alive

Nearly 2,000 years ago, a volcano named Mount Vesuvius erupted in Italy. It showered the town of Pompeii with thick layers of hot ash, burying animals and people. The rain of ash and pumice hardened around the bodies like cement.

Granite cooled slowly to form large crystals.

The soft body parts decayed and left hollow body shapes in the stone. Archaeologists later filled the hollows with plaster and dug them out of the stone.

Hard as granite

An igneous rock that turns solid underground, away from the air, is called an intrusive igneous rock. A very common example is granite. The main minerals of this hard igneous rock are quartz, mica, and feldspars.

Some columns are more than 7 ft (2 m) high.

Giant step

The Giant's Causeway is on the coast of Northern Ireland. An old legend says that giants built it as a stepping-stone pathway across the sea. But rock experts say it was made when basalt lava cooled and shrank, forming six-sided columns of solid basalt.

You'll find the minerals olivine, pyroxene, and feldspars in basalt.

Metamorphic rocks

If you've ever helped bake bread, you'll know how different the finished loaf looks from the dough. Just as dough changes into bread in the oven, some rocks change into metamorphic rocks when they heat up underground. All rocks can become metamorphic. They are changed by heat, by pressures deep in the Earth, or by both together.

Squashing snow

See for yourself how pressure and heat can change the way things look and feel. If the weather is snowy, make a snowball from light, fluffy snow. Then apply pressure by squeezing it hard. The squashed-up snowflakes melt and merge to form hard, heavy ice.

Slate splits easily into slim sheets that make good roof tiles.

Sheets and slabs

Look around almost any town, and you'll find that some houses have slate roofs. Slate was made when shale rock inside mountains was squeezed incredibly hard. Slate is used for making blackboards and floor tiles.

Marvelous marble

Marble is formed when limestone is exposed to very high temperatures deep inside the Earth. Look closely at this unpolished gray marble – its tightly packed mix of calcite minerals sparkles and glitters.

You can scratch the surface with a knife.

Marble is soft enough to carve tiny details

Mining for marble

The most famous marble comes from the Carrara quarry in Italy. The sculptor Michelangelo used it, as it was the local stone. After the marble is mined, it is sculpted or cut, and then polished to show off its beautiful colors, patterns, and shine.

Polished marble is smooth and shiny.

Small fold where the rock has bent over itself

Swirling patterns where pink granite has melted into another darker rock

Statues and sculptures

Sculptors admire marble for its beauty and strength. It can be easily sculpted without splitting.

Under pressure

This rock is called migmatite. It was formed when different rocks were squeezed and melted by heat and pressure. The melted rocks flowed and mixed together. When the mixture cooled, it hardened and became migmatite.

Black rocks

Millions of years ago, parts of the Earth were covered with dense jungles of giant ferns and swampy lagoons. As the leafy fronds (fern leaves) and stems fell onto the wet forest floor, they began to rot. Gradually, these plant remains were squashed under more layers of mud and rock. Over millions of years, the squashed plant remains became coal.

Flatter and flatter
Dead plants passed through several stages before they turned into coal. Here is what happens.

1 This is the first stage. The rotting plants did not decay completely before they were covered with mud.

2 Under the mud, the squashed plants turned into peat. Peat burns, but it is very smoky. Peat eventually becomes coal.

3 More squashing and millions of years produces anthracite, the hardest kind of coal. When it burns, it gives out the least sooty smoke.

Plant root

Deep down

For hundreds of years, people have mined (dug up) coal to use for fuel. Today we use coal to make electricity. Most coal is mined deep underground. Layers of coal, called seams, are sandwiched between layers of other rocks.

The children who pushed and dragged the carts were called hurriers.

Dangerous work

In the 19th century, young children dragged heavy carts of coal along underground passages deep inside mines. Later, ponies did this work. Nowadays, miners are highly skilled adults, but it is still a dangerous job. Miners sometimes die from the effects of poisonous gases, explosions, or collapsing mine roofs.

Children as young as five worked long hours under terrible conditions.

Diamond is the hardest natural material in the world.

Diamonds are used in jewelry. They are cut and polished so that they glitter and sparkle.

Carbon copy

What does coal have in common with a diamond and a pencil? They are both made of the same chemical – carbon. They look different because the atoms of carbon are arranged in different ways in each of them.

Another kind of carbon is called graphite. We use graphite to make the "lead" in pencils. It is so soft that it leaves marks on paper.

Sedimentary rocks

Rocks are being worn down into smaller pieces of rock and minerals by wind and water all the time. Streams and rivers wash these pieces into lakes and seas. They settle in layers as sediments that are buried and squashed together. In time, the sediments harden into new rocks, called sedimentary rocks.

A long wait

Sandstones are glued-together sand grains from beaches or deserts. But don't wait to watch them form – it may take millions of years!

Squashed pebbles stick together to make the new rock.

Pebble package

This is a conglomerate rock. It looks like a handful of pebbles cemented together. Can you see the flint pebbles? They were smoothed by water as they rolled around at the bottom of rivers or seas.

Seashells in the mountains

Some sedimentary rocks contain the remains of shells and sea creatures, called fossils. (You can read all about fossils on page 32.) Huge Earth movements lifted and moved the rocks far away from the ocean. So don't be surprised if you find a preserved seashell when you are on a mountain walk!

Layer-cake rocks

Sedimentary rocks settle in strata (layers). You can make your own strata, complete with fossils! You will need sand, plaster of paris, food coloring, a plastic bottle with the top cut off, some shells for "fossils," a bowl, a spoon, and petroleum jelly.

2 Make a layer with a different food coloring and another fossil. Repeat, building up several layers. Leave your strata to harden for a few days.

1 Put equal amounts of sand and plaster of paris and a few drops of food coloring into a bowl. Add enough water to make a smooth paste. Put the paste in the bottle. Rub jelly onto a shell and put it on top of the paste – this will be your "fossil."

The jelly covering lets the shells separate easily from the "rock."

3 Ask an adult to cut away the bottle. Break open the layers, and see your "fossils" and their imprints in their sedimentary rocks.

Larger fragments are glued together with fine particles and tiny mineral pieces.

Fragments have sharp edges.

Broken bits

Like the conglomerate on the opposite page, this breccia also contains bits of smaller rock fragments. But they have not been worn smooth. They are still sharp-edged and jagged. Some breccias are made from the broken rocks at the bottom of cliffs or on cave floors.

White rocks

Across the land and in the water, millions of creatures live in hard shells made of a mineral called calcite.

When they die, many of their shells break into tiny particles that wash into the oceans and settle on the seabed. Over thousands of years, these shells harden into sedimentary rocks called limestones.

Writing with rock
Chalk is a soft, white kind of limestone. Colored school chalk is made with dyes and minerals.

The shells are usually broken into pieces.

Shell of small mussel-like shellfish.

Chalk horse
If you scrape away the grass in a chalky area, the white rocks show through. Thousands of years ago, people cut huge horses and other shapes into chalky hillsides.

Shell sandwich
Can you spot the pieces of snail and other animal shells in this shelly limestone? They are stuck together with much tinier shell pieces – and everything is cemented together with a mineral called calcite.

Hard and soft

Limestone is very soft. Over thousands of years, rainwater dissolved a huge block of limestone in southwest China. This left the harder peaks of rock jutting out like enormous rocky eggs.

Chalk test

If you find a rock that you think is chalk, you can find out by doing a simple acid test. Chalk belongs to a group of minerals called carbonates – and all carbonates dissolve in acid. You'll need strong vinegar for the acid, a dropper, and a magnifying glass.

1 Fill the dropper from the vinegar bottle. Then carefully drip some vinegar onto the rock.

Put the rock on a glazed plate for safety.

2 Look closely at your rock with the magnifying glass. If it is chalk, it will fizz. This happens when the acid in the vinegar reacts with the chalk. It sets free some carbon and oxygen, which join into carbon dioxide bubbles.

White and straight

Next time you're in a boat, look out for white chalk cliffs on the coast. These cliffs are in Dover, England. Pounding waves have eroded the soft chalk evenly to make a straight coastline.

Waves cut into the soft chalk, and the cliff crumbles.

31

Fantastic fossils

On a rocky shore or in a stony place, you may find sedimentary rocks that look as if they were once alive. There are shapes of shells, bones and teeth, and leaves and tree bark. These are fossils – the remains of prehistoric animals and plants that were trapped in rocks and turned to stone.

Long-lost cousin
Fossils have an important job – they show us how animals and plants have changed over millions of years. The fossils you find may be very different from the animals and plants that you can see today!

Dead and buried
Swimming in the sea 400 million years ago, you might have bumped into ammonites instead of their modern relatives, octopuses and squids. Ammonites died out 65 million years ago. We know about them because their shells sank to the seabed and became fossils after the creature inside had died.

Some ammonite fossils are twice as big as you!

Shell has a spiral pattern

Direct descendant
The nearest living relative of the ammonite, the nautilus, also has a spiral shell. It swims in warm oceans and eats fish and crabs.

Around for ages

Ginkgoes are also known as maidenhair trees. People call them living fossils because the ginkgoes you see now are almost identical to those that grew 200 million years ago.

One left
Once there were dozens of ginkgo trees, but today only one kind is left.

A ginkgo leaf falls on a muddy riverbank.

How fossils form
When the soft parts of a dead animal or plant decay, the hard parts are covered in soft mud. The shape is preserved when the mud hardens into rock over millions of years.

A dinosaur dies at the same place.

The soft parts of the leaf and dinosaur rot away.

Layers of mud bury the remains.

Look for lumps that separate easily from the surrounding rock.

Is there a fossil inside?
Fossils only form in sedimentary rocks such as chalk, limestone, and sandstone. They are often inside lumps called nodules. Carefully crack these together to see what's inside.

Slowly, Earth movements and weather expose the fossils.

Cenozoic era
(65 million years ago to present)

Dinosaurs died away; mammals such as horses and cats became more numerous.

Rock record

We can learn a lot about the history of the Earth by studying rocks. The strata of sedimentary rocks have built up over millions of years. Experts look at these layers to find out which animals and plants were alive at the time. They also learn about Earth movements and about the way the climate has changed over the years.

Mesozoic era
(225 to 65 million years ago)

Long, long ago
The Earth's history is divided into three main eras (ages). Different animals and plants lived in different eras. We know how old rocks are by the kinds of fossilized animals we find in them.

Dinosaurs ruled the world, and the first birds appeared.

Paleozoic era
(600 to 225 million years ago)

Rocks from this era may contain prehistoric fish or giant tree ferns.

First fossil
In 1820, an Englishman and rock collector, Dr. Gideon Mantell, found some strange, unknown fossil teeth – and started the great dinosaur hunt!

Make an anticline

Sedimentary rock layers form one on top of another, with the oldest layer at the bottom. Earth movements can twist and fold the layers into new shapes. When layers of rock fold up to make a mountain, it is called an anticline. To make your own anticline, you will need modeling clay, a rolling pin, a knife, and some sand for "fossils."

Step back in time

The Grand Canyon in Arizona was carved by the Colorado River. It took more than 2,000 million years to form. Like a journey back in time, the rocks get older the farther down you climb.

1 Roll a lump of modeling clay into a flat strip. Sprinkle sand "fossils" onto it. This will be the oldest rock layer in your strata.

Use different colors for each different layer

2 Repeat with other colors of clay, adding more fossils between the layers. Push the two ends of the strips toward each other to make a mountain.

The youngest rocks are now on the sides of the mountain.

3 Ask an adult to slice off the top of the mountain with the knife. The layers now run sideways, from the youngest rocks to the oldest in the middle.

Worn by water

You wouldn't think that water could wear away a rock. But over millions of years, hammering raindrops, rushing rivers, and scraping glaciers rub away even the hardest of rocks. Rain also contains acids, which are strong enough to dissolve the minerals in some rocks.

Scars and scratches

Trapped under a glacier, this bit of limestone was scratched by rocks in the glacier flowing over it.

River of ice

A glacier is like a frozen river. It flows downhill very slowly, rubbing the rocks and gouging (cutting) a huge channel. You'll find glaciers in cold places such as high mountains, and near the North and South Poles.

Can you see why the tip of the glacier is called a snout?

The snout is the oldest part of the glacier.

Loose, tumbling rocks in the water rub away at fixed rocks on the river bed.

Layer of soft rock wears first

Layer of hard rock wears more slowly

Badlands

Hot days, cold nights, floods, and storms have smoothed the rocky land of the Badlands in Utah. Millions of years later, it looks like a strange lunar landscape.

River to rapid

When a river flows over rocks of different hardness, it wears away the soft ones more quickly. The slope of the water-worn rocks steepens into white-water rapids, and then into a waterfall.

Hard rock left behind forms top of waterfall.

Crashing water and boulders make a deep pool.

Smooth skippers

Try skipping pebbles next time you're near a lake or by the sea. Throw them hard and flat across the water. Well-worn, rounded ones bounce best. Sharp ones hit the water and sink.

Pebbles tossed by waves and water currents are worn smooth.

Worn by weather

When you spend the day outside, your exposed skin probably feels the effects of the weather – especially if the weather is a mixture of hot sun, cold wind, and driving rain. Rocks stay out in all weather. They are heated and windblown, wet and frozen, day after day for millions of years. When the weather cracks rocks like this, it is called weathering.

Blast that storm!
Sand whipped up by a strong wind not only gets into your beach picnic – it can blast the paint off a car! Sandblasting machines clean soot and grime from stone buildings.

Peeled like an onion
By day, the Sun heats this piece of dolerite, making it expand (get bigger). The cold night air makes it cool and contract (get smaller). Gradually the surface layers split, crack, and peel off.

Layers of stone peel away like the layers of an onion.

Dolerite is an igneous rock, which weathers easily.

Can you see how rounded the surface is? Millions of years of weathering have worn away any sharp edges.

Biggest rock
Uluru in Australia is the world's largest freestanding rock – it is over 2 miles (3.6 km) long! Weather has worn this massive block of multicolored conglomerate rock into a maze of cracks and ridges.

Crack the clay

See what happens to a rock when it warms and freezes. You'll need modeling clay, plastic wrap, water, and a freezer.

Look for cracks in the clay.

1 Moisten some clay and mold it into two lumps. Wrap them in plastic wrap. Leave one in a room, and put the other in the freezer overnight.

2 Next morning, take the ball out of the freezer. Unwrap both balls and compare them. What can you see?

Small cracks in the clay

Fading memories

If you're ever near a graveyard, take a good look at the gravestones. Over the years, plants and lichens grow on, frost bites, winds rub, and acid rain dissolves the stones. And when people clean the stones, they wear away even more of them.

3 Cracks form in the frozen ball. Put the ball back into the freezer overnight – are the cracks bigger the next day?

Look at the dates. The oldest stones are usually the most worn.

Icicles of rock

Not all the rocks under your feet are solid. In areas with a lot of limestone, rainwater and streams dissolve away the rock. Over thousands of years, small cracks become tunnels. Drips on the roof and floor build pointed "icicles" of rock called stalactites and stalagmites. It is a pitch-black wonderland – until the cave explorer's light shines.

Roof hanger
Look out for wet stalactites hanging from cave roofs. If you shine a torch at them, they will glisten.

Cave country
When rainwater seeps through the soil, it becomes a weak acid called carbonic acid. Carbonic acid eventually dissolves limestone. Look for small holes in limestone areas. They may be clues to a world of underground caves.

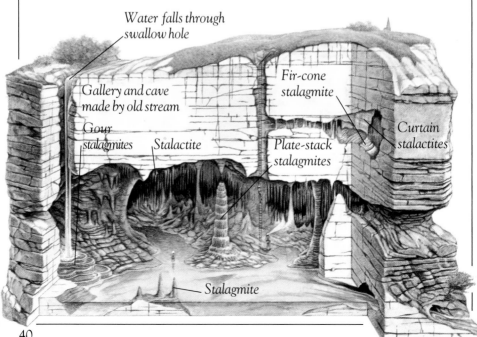

Water falls through swallow hole

Gallery and cave made by old stream

Fir-cone stalagmite

Gour stalagmites

Stalactite

Plate-stack stalagmites

Curtain stalactites

Stalagmite

Instant stalactite

Why not make your own stalactites in just a few days? You will need a piece of wool, paper clips, a jug, a dish, a spoon, two glass jars, and baking soda.

1 Spoon baking soda into two jars of warm water. Stir and repeat until no more soda dissolves.

2 Put each end of the piece of wool, weighted down with paperclips, into each jar. Place a saucer between them.

3 The solution seeps along the wool and drips onto the saucer.

Baking soda stalactite

Baking soda stalagmite

Stalactites

Water dissolves a mineral called calcite from limestone. Each drop of water leaves a tiny bit of calcite, which gradually builds into a pointed shape that hangs down.

Stalagmites

Stalagmites also grow from water drips. They form on the floor of the cave and grow upward as the drips evaporate and leave calcite deposits.

The tip is the newest part of each stalagmite.

Slow grower

Some stalactites only grow as wide as your finger every 200 years. This means that some large specimens are more than 10,000 years old!

Crystal collection

Most crystals form when heat deep inside the Earth melts minerals in rocks. Later, the minerals cool and harden into new structures called crystals. Crystals are evenly shaped because they are made of tiny atoms that arrange themselves in the same regular way as the crystal grows.

Calcite crystals

Quartz crystals

Each pisolite crystal has a light centre and darker layers.

Crystal ball
Different crystals grow into different shapes. Here, calcite crystals have grown into small round shapes, while the jagged, pointed crystals are formed by clear quartz.

Onion skins
Pisolites are small, ball-shaped groups of crystals. They grow in layers just like the layers on an onion. This limestone has been cut open and polished, so you can see the pisolites inside it.

Grow your own crystals

A crystal grows by adding layers of its mineral to its surface. You can grow your own crystals with salt, warm water, a jar, a pencil, and wool.

1 Dissolve as much salt as possible into a jar of warm water. Keep adding and stirring until you can see undissolved salt crystals in the water.

Like a tree

These dendritic copper crystals come from Australia. Dendritic means "tree-like." This crystal looks like it has branches, twigs, and even tiny leaves!

Crystals are long and slim, like the branches of a tree.

2 Tie the wool around the pencil so it rests in the salt water. Leave the jar in a cool place.

3 As the water cools, salt crystals grow on the wool. Each day, pour the water away. Add a fresh batch full of dissolved salt. Day by day, the crystals on the string get bigger.

Bigger crystals take longer to grow than smaller ones do.

Kettle crystals

You may find crystals in a teakettle. Minerals are often dissolved in certain kinds of tap water. These crystallize and coat the inside of the teakettle when the water is boiled.

Needle-like crystals from the inside of a teakettle

Don't let your kettle boil dry. The fur grows extra-fast!

Hard and soft

As you explore the world of rocks and minerals, you can use certain features to identify them, such as their color and hardness. Like the experts, you can measure a mineral's hardness by trying to scratch it with various things. The hardness scale goes from 1 to 10, with the softest minerals at 1.

Soft as a baby
The softest mineral is talc – look for it in your bathroom! Raw talc looks like a pearly-white rock, but it is very soft, light, and slippery.

Don't touch film – the grease on your fingertips will ruin it!

Some old telescopes have clear fluorite lenses.

Photographic film contains tiny crystals of silver.

Medium minerals
Most minerals are medium-hard. You may see silver (hardness 3) in precious jewelry. But most silver is used in camera film. Clear crystals of fluorite (4) are used in some telescopes and microscopes.

How hard is soft?
Each mineral has a certain hardness. To measure it, a German geologist named Friedrich Mohs invented the Mohs' scale in 1822. He used ten minerals; the softest is talc (1), the hardest is diamond (10). A mineral on the scale is scratched by those above it, but scratches those below it.

Talc – this is the softest mineral

Gypsum

Calcite

Fluorite

1   2   3   4

Hard minerals

Take a close look at a wind-up watch. It may say "jeweled movement," or the number of jewels. Parts with hard jewels in them last for years. A "quartz" watch contains a tiny quartz crystal (7) that vibrates and keeps accurate time.

These ruby crystals are artificial (grown by people).

Jade jacket

The ancient Chinese believed that jade (6-7) would make people live forever; 2,000 years ago, they made funeral suits from jade, to bury the rich and famous.

This suit once belonged to a princess. It is jade linked with gold.

Green as grass

Emerald (7.5), is a brilliant green form of the mineral beryl. Emeralds are among the most valuable of all gems, because they are extremely hard as well as rare and beautiful.

Tiny amounts of chromium give the green color.

This kind of jade is called nephrite.

Scratch it yourself

Try out Mohs' scale for yourself. If you can scratch an object with your fingernail, it has a hardness of less than 2.5. An object you can scratch with a copper coin has a hardness of less than 3.

Apatite

Orthoclase feldspar

Quartz

Topaz

Corundum

Diamond – this is the hardest mineral

| 5 | 6 | 7 | 8 | 9 | 10 |

Uncovering gemstones

You want to give someone a precious gift. If you had plenty of money, what would you choose? Jewels and gems are valuable because they are beautiful, rare, long-lasting, and treasured. But first they must be found. They can be collected at the surface of the Earth or mined from deep underground. Finally, they are cut and polished to make them sparkle.

Glistening colors change with light.

Silica grains reflect and scatter light into rainbow colors.

Cool down under
The town of Coober Pedy in South Australia is famous for its opals. It is extremely hot above ground during the day, but is much cooler down inside the mine, so mine holes near the surface are turned into cool homes for the miners.

Uncut block of black opal

Flashing opals
In ancient Rome, opals meant power. In the 15th century, they were considered unlucky. Opal is a semi-precious gem. It is unusual because it has no crystals and is made of tiny balls of silica.

Sapphire has grown together with crystals of hard, glassy, spinel.

Emery is corundum mixed with other minerals. It is used to make nail files, like this one.

Corundum colors

Although rubies are red and most sapphires are blue, they are both forms of the mineral corundum (aluminum oxide). The blue color in the sapphire comes from tiny amounts of iron and titanium. The red color of the ruby comes from the metal chromium.

Red ruby

This famous ruby is called the Edwardes Ruby. The most valuable rubies are this deep color, known as "pigeon's blood red."

Small pebbles roll along the river bed, becoming round and smooth.

Here are the diamonds!

Spot the gems

You don't just find diamonds in diamond mines. Diamond crystals are sometimes found in river gravel, because they are hard enough to stand the tumbling and wear of the water. There are three diamonds in this pile of gravel – can you spot them?

Little reward

Diamond is valuable because it is very hard, and also because it is rare. You would have to dig through enough rock to fill a school classroom to find one pea-sized diamond!

Precious gems

Finding a dull-looking gemstone among a mass of rock is only the first step to a glittering jewel. An expert called a lapidary carefully cuts and polishes the stone, gradually turning it into an object of great value and beauty. Each type of gemstone has a traditional cut shape that shows off its sparkle and color.

This diamond is embedded in a kind of rock called kimberlite.

Raw diamond

Diamonds are formed when pure carbon is crystallized (made into crystals) under enormous pressure and heat deep inside the Earth. Many diamonds are mined in the Kimberley area of South Africa.

Diamond cutting

Follow the steps to see how the diamond above is changed into a beautiful gem.

1 Foundation
The lapidary grinds the crystal round and cuts off the top point.

2 Facets
The next step is to grind flat patches (called facets) at exact angles.

3 Finished
Fifty-eight facets reflect the light, giving the diamond its fire (sparkle).

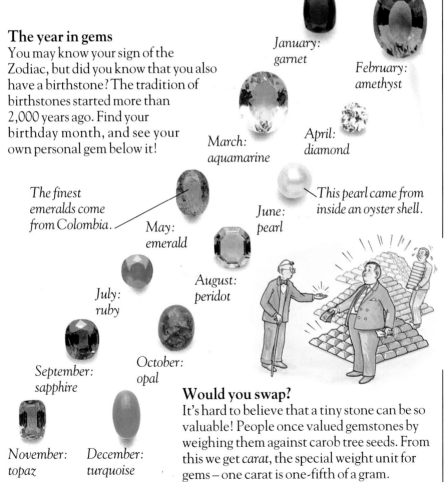

Smoky-yellow topaz is the rarest.

Hard but brittle

Topaz makes beautiful jewels. It is extremely hard (8 on Mohs' scale), but it is also brittle. It cracks or splits easily in a certain direction.

Topaz comes in many different colors.

The year in gems

You may know your sign of the Zodiac, but did you know that you also have a birthstone? The tradition of birthstones started more than 2,000 years ago. Find your birthday month, and see your own personal gem below it!

January: garnet

February: amethyst

March: aquamarine

April: diamond

The finest emeralds come from Colombia.

May: emerald

June: pearl

This pearl came from inside an oyster shell.

July: ruby

August: peridot

September: sapphire

October: opal

November: topaz

December: turquoise

Would you swap?

It's hard to believe that a tiny stone can be so valuable! People once valued gemstones by weighing them against carob tree seeds. From this we get *carat*, the special weight unit for gems – one carat is one-fifth of a gram.

Marvelous metals

Can you imagine life without metals? There would be no steel for cars, and no silver and gold for coins and jewelry. Most metal objects are made of alloys (mixtures of metals). Metals come from rocks called ores. Tiny pieces of metal are scattered through the ore, and the ore is crushed to collect the pure metal.

Hematite is also called kidney ore. Can you see why?

Main ore
Hematite is the main ore for iron. It is dug out of huge mines and taken to ironworks, where it is crushed and made ready for smelting (melting).

Red-hot and runny
To collect pure iron, iron ore is mixed with coke (a type of coal) and limestone. Then it is melted in a giant oven called a blast furnace. Red-hot runny iron forms in the furnace and pours into huge containers known as ladles.

Some blast furnaces can work non-stop for ten years.

Iron items
Mixing iron with tiny amounts of carbon makes steel. Pure iron is also hammered or cast into many things, including wrought-iron railings.

Going for the gold

If you were hunting for gold, you would be very lucky to find a nugget of it – or even a gold crust on another mineral. Usually, tiny specks of gold are mixed into ores.

Crust of gold on quartz

This ring is made from white-, red-, and yellow-colored gold.

Quartz

Chalcopyrite is the main ore for copper.

Don't be fooled

This chalcopyrite rock seems to glisten with gold specks. But the specks are a less-valuable mineral than gold, which is why chalcopyrite and pyrite are called "fool's gold." Old-time miners often thought they had found a fortune!

✋ *Be careful near running water.*

Try not to scoop up too much water, or you might get wet when you swirl the pan.

Panning for gold

Some river sand and gravel has tiny gold grains in it. Using an old frying pan, try scooping up a panful from a shallow stream and swirling it around. When the water has washed the sand away, look for the heavier gold in the bottom of the pan.

Golden fleece

In ancient times, people used sheep's wool to filter gold flecks from rivers – although it is unlikely that they used a whole sheep!

Rocks and soil

Look at the ground in a garden or park. Can you see how bare patches of rock have almost no plants on them? But solid rock breaks down, along with rotting plants and animals. This mixture becomes soil – and so even the hardest, barest rock becomes a home for living things.

Big bloomers
Fertilizers contain extra nutrients (foods) that might not be in the soil. Plants fed on the right fertilizer grow big, strong, and healthy!

Feet of clay
Have you ever walked across a springy patch of moss, or across a heavy clay field where the soil sticks to your shoes? There are dozens of different kinds of soils, and soil covers most of the world's land. Each type of soil is different. It varies with the rocks beneath, the local rainfall and climate, and the plants growing there.

The grains in dry sand soil are 100 times larger than clay particles.

If you pick up a handful of clay soil and mash it into a lump, it will stick together.

Chalk soil is light and powdery. Water quickly drains through it.

Peat soil is dark and moist, with matted fibers and bits of partly rotted plants.

See what's in the soil

Finding out about the soils in your area could help you become a prize-winning gardener! Make a soil survey by collecting soil samples from different areas. By separating the ingredients of each sample, you can compare the different soils. You will also need a trowel, screw-top jars, and some water.

Organic matter floats

1 Use plastic bags to gather different soils. Put two handfuls of each sample into separate jars.

2 Add water. Screw on the lid and shake hard for a few minutes. Leave overnight and compare your samples the next day.

Heavy mineral particles sink and may form layers.

Back to the Earth

Make your own natural fertilizer! Pile weeds and vegetable peelings into a big heap called a compost heap. They will rot into a rich compost. You can then spread the compost over flower and vegetable beds to enrich the soil.

Layers in the soil

Most soils have several layers. On the surface is humus – fragments of rotting leaves and other once-living things. Below is topsoil with plant roots and worms. Under this is subsoil, then crumbled rocks, and finally solid bedrock.

Upper roots take in water and nutrients from the topsoil.

By the sea

Have you ever spent a summer day playing on the warm sand and sun-baked rocks by the sea? The same beach looks very different during a winter storm. Waves crash, stones tumble, winds blow, and rains batter the rocks. Over hundreds of years, rocky cliffs become fine silvery sand, and powerful waves eat away at the land.

White sand is made from ground quartz, seashells, and coral.

Gold sand has mica grains in it.

Gray sand usually comes from rocks that contain granite and feldspar.

Only rocks

On the coast, rocks and minerals are everywhere you look. This is because wind and waves wash away most soil and plants, leaving only bare rocks.

The sea slowly cuts into the cliff. Lumps of cliff fall off into piles of sharp boulders.

Mini-rocks

Sand is made from tiny worn-down rocks called grains. Its color depends on the original rocks and minerals. A black sandy beach may have once been the dark lava of a volcano.

The falling tide leaves a line of debris on the shore.

_Coarse
(medium)
pebbles_

_The waves swish
away the grains, leaving
the larger, heavier pebbles._

Sort them out

Like a sieve, the sea sorts out small sand grains from bigger stones. See how it works by gathering a mixed bucketful of grains and pebbles. Dump them near the waves, and watch what happens.

_Small
pebbles_

Moving house

On some coasts, the sea carves away yards of land each year. Lighthouses that warn ships of dangerous rocks sometimes have to be moved inland as the cliff edge gets nearer.

_Finest
pebbles_

_Sand
grains_

_Winds blow from sea
to shore, piling up sand
behind the beach into
sand dunes._

_Groynes (wooden
fences) stop sand from
sweeping out to sea._

Rounder and smaller

The sea crashes loose rocks against each other. Their sharp edges wear off and they gradually become smaller – until they are just bits of sand, silt, and mud.

_These cliffs will
eventually wear
away into the sea._

In the desert

Scorching hot days, bitterly cold nights, and wind-blown dust and sand shape a dramatic desert landscape. Deserts can be sandy, pebbly, or stony. Sand blown by the wind is just like sandpaper. It wears away soft rock and leaves harder rock layers sticking out. Temperature changes and water erode the landscape and break up rocks into fantastic shapes.

Flat-topped area called a plateau ends sharply at steep cliff

Hands of stone
These sandstones are in Monument Valley, Utah. Can you see why they are called The Mittens?

Hot sun and cold nights make dew-damp rocks crack and break up.

Softer rock wears away, leaving an arch.

An inselberg is a rounded hump of harder rock.

Blowing sand wears the base of a large stone, leaving a shape called a pedestal rock.

Red color of sandstone comes from iron minerals

Compare a rock on the top of the pile with one from the center. Can you feel a difference?

Sand grains may become solid sandstone.

Rock solid

About 200 million years ago, these sand grains were blowing across a desert. The constant rolling and bouncing made them round. Over the years, the sand grains squashed together and cemented into solid sandstone.

Sun shelter

Rocks shatter when they absorb the sun's heat. But this helps animals, which find a cool home in the shade underneath. Pile rocks in the hot sun, and feel them after a few hours. Have the shady rocks stayed cool?

A small, separated plateau is called a mesa

Follow the wind arrows – the wind blows the sand into hills called dunes.

Sand to soil

An oasis is an area in which rocks hold water and make a pool or well. Plants may grow near it. Like little anchors, the plant roots hold the sand grains together. Gradually, the shifting sand dunes become stable soil.

Water is carried to the oasis in rock layers under the surface.

Plants grow in the oasis if their roots can reach water.

Building with rocks

Rocks have been used to construct buildings for thousands of years. The ancient Romans carved up stones to build roads that we still use today. Nowadays, people build with bricks, tiles, and concrete. These modern materials are easy to cut and shape and are cheaper than quarrying stone.

Sails on the shore

Sydney Opera House in Australia, is a famous modern building. It looks like a ship with billowing sails – but these sails are made of concrete, covered with gleaming white tiles.

Concrete is a mixture of cement, sand, gravel, and water.

Concrete buildings are inexpensive to build.

This townhouse is made of bricks.

Paint on walls helps stop weather wearing away the stones

Wooden frame is filled with clay bricks

Stone survey

Take a good look at the buildings in your town. Use your notebook to sketch walls, buildings, and statues. Find out if they are made of natural stone or manufactured bricks and concrete.

Roman villa

Rich Romans had elegant houses decorated with beautiful stone statues and floors. The builders used local rocks, cut into blocks. For columns and statues, they carried marble over thousands of miles.

Terra-cotta roof

Oolitic limestone

This sandstone is called Old Red Sandstone.

This limestone can be crushed and heated with clay to make cement.

Timber-frame house has a solid sandstone base

Modern terra-cotta roof tile

Terra-cotta tile

The word terra-cotta means "baked earth." It is a kind of clay that has been baked in a special oven called a kiln. Through the ages, clay has been shaped into jars, pots, bricks, tiles, and ornaments. The ancient Greeks made beautiful terra-cotta objects.

Carry maps and information leaflets from the local museum in your backpack.

Index

A

alloys, 50
amber, 11
amethyst, 49
anthracite, 26
apatite, 45
aquamarine, 49
ash, 23

B

basalt, 22, 23;
    olivine, 17
beryl, 45
birthstones, 49
breccia, 29

Diamonds in gravel

C

calcite, 25, 30, 41, 42, 44
carat, 49
carbon, 27, 48, 50
carbonates, 31
carbonic acid, 40
caves, 40
chalcopyrite, 51
chalk, 30, 31, 52
clay, 52, 59
coal, 26-27
concrete, 58
conglomerate rock, 28
coral, 11
corundum, 45, 47
crust, 16, 18, 28, 35
crystals, 8, 22, 42-43

DE

desert, 56-57
diamond, 27, 45, 47, 49;
    cutting, 48
dolerite, 38

Insect in amber

Earth;
    history, 34;
    inside the, 16-17
eclogite, 17
emerald, 45, 49
emery, 47
equipment, 9
erosion, 36-37, 38-39, 54, 55, 56

Pompeii figure

FG

feldspar, 10, 23, 45
fluorite, 44
fool's gold, 51
fossils, 28, 29, 32-33

gabbro, 12
garnet, 17, 49
gemstones, 46-47, 48-49

geologists, 9
glacier, 36
gneiss, 12
gold, 51
granite, 10, 23
graphite, 27
gypsum, 44

I

ice, 24, 36
igneous rocks, 12, 22-23
iron, 47, 50

Smoky-yellow topaz

JL

jade, 45
jet, 11
jewels, 45, 48

lapidary, 48
lava, 19, 21;
    ropy, 20
    frothy, 22
limestone, 25, 31, 40, 41;
    oolitic, 59
    shelly, 30

M

magma, 20, 21
Mantell, Dr. Gideon, 34
mantle, 16, 17
marble, 25
metals, 50-51

*Sea washing
away sand*

OPQ

RS

Granite

Earthquake

TVW

Acknowledgments

**Dorling Kindersley
would like to thank:**
Redland Roof Tiles for
supplying a terra-cotta tile
for photography.
Lynn Bresler for the index.

Illustrations by:
Peter Visscher,
Nick Hewetson,
Raymond Turvey.

Picture credits
t=top b=bottom c=center
l=left r=right
Bruce Coleman Ltd:19tr,
39b; /David Austen 9tl;/ Bob
& Clara Calhoun 37tr;/ John
Cancalosi 40t; /M. Fogden
21c; /Jeff Foot 35l; /Prato
18br.
Andreas von Einsiedel: 5, 8br,
20t, 20c, 23tr, 24b, 25b, 28c,
29b, 30b.
Robert Harding Picture
Library: 23b, 36b, 45cr, 58c;/
A. Waltham 31t.
Colin Keates: 4tr, 8cl, 10,
11bc, 17t, 17c, 17bl, 22b, 36t,
38c, 44-45b, 45t, 46b, 47tl,
47bl, 48c, 48b, 49t, 49c, 50c,
51tc, 51c, 57tl, 59cl.
Science Photo Library: Peter
Menzel 46c; /Alfred Pasieka
11tl; /Heine Scheebeli 50bc.
Zefa: 11tr, 19c, 20b, 31b,
41r, 56c; /Hummel 38b;/
J. Schorken 25c.